KRAKEN

HYDRA

Disclaimer:

The creatures in this book are not real. They are from myths. They are fun to imagine. Read the 45th Parallel Press series Magic, Myth, and Mystery to learn more about them.

Published in the United States of America by Cherry Lake Publishing
Ann Arbor, Michigan
www.cherrylakepublishing.com

Reading Adviser: Marla Conn, MS, Ed., Literacy specialist, Read-Ability Inc.
Book Designer: Melinda Millward

Photo Credits: © SilviaP_Design /pixabay.com, back cover, 16; © Joe Therasakdhi/Shutterstock.com, cover, 5, 10; © Christophe Kiciak/Alamy Stock Photo, cover, 5; © Unholy Vault Designs/Shutterstock.com, 6; © HWitte/Shutterstock.com, 9; © fotokostic/istockphoto.com, 12; © delcarmat/Shutterstock.com, 15; © dvukkostic91/depositphotos.com, 19; © GraphicsRF/Shutterstock.com, 19, 20; © vectortatu/Shutterstock.com, 20; © Christianm/Dreamstime.com, 21; © NNNMMM/Shutterstock.com, 23; © Kazakova Maryia/Shutterstock.com, 24; © Leonardo Gonzalez/Shutterstock.com, 25; © Yulia Mayorova/Shutterstock.com, 25; © fantasticpicture/pixabay.com, 27; © Claudia Prommegger /stock.adobe.com, 29

Graphic Element Credits: © studiostoks/Shutterstock.com, back cover, multiple interior pages; © infostocker/Shutterstock.com, back cover, multiple interior pages; © mxbfilms/Shutterstock.com, front cover; © MF production/Shutterstock.com, front cover, multiple interior pages; © AldanNi/Shutterstock.com, front cover, multiple interior pages; © Andrii Symonenko/Shutterstock.com, front cover, multiple interior pages; © acidmit/Shutterstock.com, front cover, multiple interior pages; © manop/Shutterstock.com, multiple interior pages; © Lina Kalina/Shutterstock.com, multiple interior pages; © mejorana/Shutterstock.com, multiple interior pages; © NoraVector/Shutterstock.com, multiple interior pages; © Smirnov Viacheslav/Shutterstock.com, multiple interior pages; © Piotr Urakau/Shutterstock.com, multiple interior pages; © IMOGI graphics/Shutterstock.com, multiple interior pages; © jirawat phueksriphan/Shutterstock.com, multiple interior pages

45th Parallel Press is an imprint of Cherry Lake Publishing.

Library of Congress Cataloging-in-Publication Data

Names: Loh-Hagan, Virginia, author.
Title: Kraken vs. Hydra / by Virginia Loh-Hagan.
Other titles: Kraken versus Hydra
Description: Ann Arbor, Michigan : Cherry Lake Publishing, 2020. | Series: Battle royale : lethal warriors | Includes index.
Identifiers: LCCN 2019032921 | ISBN 9781534159389 (hardcover) | ISBN 9781534161689 (paperback) | ISBN 9781534160538 (pdf) | ISBN 9781534162839 (ebook)
Subjects: LCSH: Kraken–Juvenile literature. | Hydra (Greek mythology)–Juvenile literature.
Classification: LCC QL89.2.K73 L643 2020 | DDC 001.944–dc23
LC record available at https://lccn.loc.gov/2019032921

Printed in the United States of America
Corporate Graphics

About the Author

Dr. Virginia Loh-Hagan is an author, university professor, and former classroom teacher. She wrote 45th Parallel Press books about Hydra and Kraken. Her favorite animals are dragons, octopus, and jellyfish. She lives in San Diego with her very tall husband and very naughty dogs. To learn more about her, visit www.virginialoh.com.

Table of Contents

Introduction

Imagine a battle between the Kraken and Hydra. Who would win? Who would lose?

Enter the world of *Battle Royale: Lethal* **Warriors**! Warriors are fighters. This is a fight to the death! The last team standing is the **victor**! Victors are winners. They get to live.

Opponents are fighters who compete against each other. They challenge each other. They fight with everything they've got. They use weapons. They use their special skills. They use their powers.

They're not fighting for prizes. They're not fighting for honor. They're fighting for their lives. Victory is their only option.

Let the games begin!

VS.

KRAKEN

The Kraken's tentacles can reach the top of tall ships.

The Kraken is a sea monster from **Norse** myths. Norse means from the area around Norway. It lives a long time because nobody can kill it. It's half-squid. It's half-crab. It has a scary **maw**. Maw is a beast's mouth and jaws. Its maw has many rows of teeth. The teeth are 5 feet (1.5 meters) long. Its maw can open really wide. It has a beak. The beak is sharp. It can shatter anything.

The Kraken is over 1 mile (1.6 kilometers) long. It's the size of 10 ships. It has 12 **tentacles**. Tentacles are long, moving arms. It has over 1,000 suckers. The suckers have tiny teeth. The tentacles have big claws.

The Kraken lives in deep northern oceans. It lives in caves. It uses its tentacles to **anchor** itself. It sticks to the ground. It crawls along the ocean floor.

The Kraken hunts at night. It eats anything that comes by. It lays low. But sometimes it surfaces. It attacks ships. It wraps its big tentacles around ships. It **capsizes** ships. Capsize means to turn over. Sailors fall out. The Kraken's smaller tentacles grab sailors. The Kraken uses its beak to eat sailors. It stuffs sailors into its maw.

The Kraken can see in total darkness.
It can see far away.

The Kraken sneaks up. It attacks. It goes back underwater. This makes a **whirlpool**. Whirlpools are powerful pools of swirling water. They have the strength of hurricanes. They trap and sink ships.

The Kraken has many tricks. It can **camouflage** itself. It blends into its surroundings. The Kraken moves at a high speed. It uses jets of water. It launches itself through water. It's a powerful swimmer. It leaves behind black oil. This oil blinds prey. Its spit has poison. It doesn't have bones. It pulls itself through tight spots.

Sometimes, it loses body parts during an attack. But it can regrow body parts. The new parts are soft. This makes the Kraken weak. So, it hides until it's strong again.

FUN FACTS ABOUT THE KRAKEN

- The Kraken has been featured in several stories. Jules Verne wrote about the Kraken in *20,000 Leagues Under the Sea*. In the book, a big squid attacks the submarine. Alfred Lord Tennyson was a famous poet. He wrote a famous poem called "The Kraken."
- Ichthyosaurs were fish lizards. They are extinct. Extinct means they no longer exist. Scientists found Ichthyosaur bones. An Ichthyosaur had been squeezed by a large tentacle. Some thought this was proof of a giant squid or the Kraken. Ichthyosaurs were 30 feet (9 m) long. It would take a large sea monster to eat them.
- In 1782, 10 British warships disappeared. Pierre Denys de Montfort studied animals. He blamed the Kraken for the missing ships. A survivor shared the truth. The ships had been lost in a hurricane. Montfort's reputation was ruined.
- *Daedalus* was the name of a ship. It encountered the Kraken in 1848. Sailors said it was over 60 feet (18 m) long. Sir Richard Owen said they saw a seal. This caused a big fight between Owen and the captain of the *Daedalus*.

HYDRA

Sometimes, the Hydra is called the Lernaean Hydra.

The Hydra is a sea monster from Greek myths. It lives in Lake Lerna. Lerna was in ancient Greece. It's like a swamp. It's a gate to the **underworld**. The underworld is a place where dead souls live.

The Hydra is a powerful dragon. It guards the underworld. It lives in an underwater cave. It stays close to the lake. But at times, it leaves to eat. It goes into town. It hunts. It kills. It eats animals. It eats people. It was born to be a killer.

The Hydra has a snake's body. It has 2 arms. It has 2 legs. It has 9 heads. It has one main head. It's in the middle. This head is **immortal**. This means the head can never die. It can live forever. It lives after being cut off. It can't be harmed.

Each head has many sharp teeth. It can bite. It can tear. It can rip. Each head also has sharp horns. It can ram. It can punch. Some heads spit magical water. It can blast out its enemies. It can send its victims flying. It also uses its long necks. It lunges. It attacks from different positions.

Ancient Greek painters couldn't fit all of the Hydra's heads on a vase.

The Hydra has sharp claws. It has sharp spikes on its spines. It has a long tail. Its blood has **venom**. Venom is poison that has to be injected. The Hydra has sharp **fangs**. Fangs are pointed teeth. It injects poison by biting. The poison causes much pain before death. Even the Hydra's breath is poisonous. It's also stinky. Just smelling its breath could be deadly. Its spit is poisonous. It spits **acid**. Acid burns through things. It destroys things.

The Hydra is really hard to kill. It has super healing powers. It can regrow its heads. It grows 2 heads for every 1 lost head. This happens quickly. The Hydra can keep attacking.

FUN FACTS ABOUT THE HYDRA

- Hera was a Greek goddess. She was the queen of gods. She turned the Hydra into a constellation. Hydra is the largest constellation. It's also the longest. It looks like a water snake. It's located in the area of the sky known as the Sea.
- "Hydra" is also a real word. It means a hard problem. Hydra problems are hard to solve. When one problem is solved, other problems occur. Hydra problems never end.
- Albertus Seba collected animals. He called it his "cabinet of curiosities." He wrote a book. He included a picture of "the Hamburg Hydra." Carl Linnaeus proved it was fake. The Hydra was made from weasel heads and snake skins.
- Echidna is the Hydra's mother. She's the "mother of monsters." She's half-woman. She's half-snake. She has snakes for hair. Typhon is the Hydra's father. He's the "father of monsters." He's a giant. His legs are snake coils. He hisses as he moves. He has glowing red eyes. He has 100 dragon heads. The heads spit fire. They make a lot of noise. He has hundreds of wings. His fingers are dead snakes.

CHOOSE YOUR BATTLEGROUND

The Kraken and Hydra are fierce fighters. They're well-matched. They both have similar powers. They both have the power to regrow damaged body parts. But they have different ways of fighting. They also have different weaknesses. So, choose your battleground carefully!

Battleground #1: Sea

- The Kraken lives on the ocean floor. It knows how to live deep in the ocean. It mainly lives around Norway, Iceland, and Greenland. These countries are surrounded by water.

- The Hydra lives underwater. It spends its life by water. Lerna has a lot of water springs. The Hydra is a type of sea **serpent**. Serpents are snakes.

In some stories, the Hydra has thousands of heads.

Battleground #2: Land

• The Kraken can't leave the water. It can only live in the sea. It comes up for air at times. But it can't really breathe on land. It would die on land.

• The Hydra prefers the water. But it can fight on land. It can breathe on land and underwater.

Battleground #3: Mountains

• The Kraken can't leave the water. Mountains are on land. It would die if it was out of water for too long.

• The Hydra can fight on mountains. It lives in a cave. It hides behind big rocks. Lerna is between the mountains and ocean.

ARMED AND DANGEROUS: WEAPONS

Kraken: The Kraken's best weapon is its huge size. It can take over anything and everything. It can create deadly whirlpools. They're formed when 2 opposing currents meet. Currents are waves of water moving in the same direction. A whirlpool that sucks things underwater is called a vortex. It pulls things to the bottom of the sea.

Hydra: Most living things regenerate. Regenerate means to grow again. It means to heal from damage. When living things get cut, they can heal. Tissues and organs renew. Scars form. Hair and nails grow. But the Hydra has superpowers. Its heads grow on top of cut stumps. It quickly re-creates lost or damaged parts. Its cells change. It regenerates. The Hydra is very healthy. It doesn't get sick. It doesn't age.

FIGHT ON!

The battle begins! The Kraken and Hydra fight in the sea. Both the Hydra and Kraken are comfortable with the sea. They're strong swimmers. They know how to handle sea storms.

Move 1:

The Hydra looks for large **schools** of fish. Schools are groups of fish. The Hydra looks for bubbling waters. These are signs of the Kraken. The Hydra knows the Kraken likes to sneak up on opponents. It keeps a close eye on the waters. When it sees the Kraken, it knocks the water around. It wants the Kraken to come to the surface.

Hercules was tasked with killing the Hydra.

Move 2:

The Kraken slowly rises to the top. It covers its nose and mouth with a tentacle. It protects itself from the Hydra's poisonous breath and smell.

Move 3:

The Hydra sees one of the Kraken's tentacles. Its main head breathes out fire. The Kraken's tentacle gets burned off.

Move 4:

The Kraken begins the process of regrowing its tentacle. Another tentacle grabs one of the Hydra's heads. Its suckers squeeze the head off. But some of the Hydra's blood gets on the Kraken. The Kraken quickly bites off its own tentacle. It does this so the poison won't spread.

In some stories, heroes dip weapons in the Hydra's blood.

LIFE SOURCE: FOOD FOR BATTLE

Kraken: Since the Kraken lives in the sea, it mainly eats fish. The Kraken eats a lot of fish. It can eat all kinds of fish and sea creatures. Its beak is able to break through hard shells. It can swallow some fish whole. It also eats sailors who fall into the ocean.

Hydra: The Hydra likes to eat human flesh. The Hydra wasn't born eating humans. But once it tasted human meat, that's all it would eat. The Hydra adapted to its living space. Its food sources are fish and sailors. Sailors are easier to hunt in the water. They can't swim or run away. The Hydra needs to eat fish as well. Humans aren't that nutritious. They don't provide enough energy.

Move 5:

The Hydra's head grows back. It grows 2 heads. Its head regrows faster than the Kraken's tentacles.

Move 6:

The Kraken's tentacles attack the Hydra's heads. The Kraken is much bigger than the Hydra. It's maw can break the Hydra's bones.

Move 7:

The Hydra protects its main head. The Kraken wants to cut it off. It plans on taking the head to the ocean floor. It wants to put a rock over it.

Move 8:

The Kraken slowly circles around the Hydra. It wants to trap the Hydra in a whirlpool.

Monster hunters have tried to trap the Kraken. They've failed.

AND THE VICTOR IS . . .

What are their next moves?

Who do you think would win?

Kraken could win if:

- It cuts off the Hydra's heads and sets fire to the stumps. This is so the heads can't grow back. But this has to be done quickly.
- It could get strong sailors to distract the Hydra. Sailors could keep chopping off the Hydra's many heads. While the Hydra is distracted, the Kraken can attack.

Hydra could win if:

- It gets the Kraken on land. It's used to being on the bottom of the ocean. It won't survive long on land.
- It avoids being around tall ships and whirlpools. The Kraken likes to do surprise attacks. The Hydra needs to always be alert.

Both the Kraken and Hydra are feared monsters in ancient myths.

Kraken: Top Champion

Like the Kraken, Scylla and Charybdis were scary sea monsters. They were from Greek myths. They were born women. Then, they were turned into monsters. They lived on opposite sides of the Strait of Messina. A strait connects two bodies of water. The Strait of Messina is in Italy. Scylla represented the dangers of the rocky shore. She lived under a big rock. She had 12 feet. She had 6 long necks. She had 6 heads with sharp teeth. She had a belt of barking dog heads. She hid in a sea cave. She kidnapped sailors from their ships. She ate them alive. Charybdis represented a deadly whirlpool. She lived under a small rock. She drank the waters. She swallowed ships and sailors. She also swallowed islands whole.

Hydra: Top Champion

Yamata no Orochi was a Japanese Hydra. It had 8 heads. It had 8 tails. It had red eyes. It had trees growing on its back. It was shaped like a giant snake. It was as big as 8 valleys and 8 hills. It liked to eat pretty maidens. Maidens are young women. Susanoo was a storm god. He tricked his sister. His sister was the sun goddess. Susanoo was kicked out of heaven for doing this. He met a couple. The couple was crying. They had to give one of their daughters to Orochi every 7 years. This made them sad. They didn't want to give up any more daughters. Susanoo helped them. He tricked Orochi by making him sleepy. Then, he cut Orochi into pieces. Inside of Orochi, there was a great sword. Susanoo gave the sword to the sun goddess.

Consider This!

THINK ABOUT IT!

- How are the Kraken and Hydra alike? How are they different? Are they more alike or different? Why do you think so?
- What skills do you have to fight the Kraken? What skills do you have to fight the Hydra?
- Would you rather be huge like the Kraken or regenerate like the Hydra? Explain your choice.
- Learn more about Greek and Norse mythological monsters. How do the Kraken and Hydra compare to other monsters?
- Read the 45th Parallel Press books about the Kraken and Hydra. What more did you learn about these monsters?

LEARN MORE!

- Loh-Hagan, Virginia. *Hydra*. Ann Arbor, MI: Cherry Lake Publishing, 2016.
- Loh-Hagan, Virginia. *Kraken*. Ann Arbor, MI: Cherry Lake Publishing, 2017.
- McKerley, Jennifer Guess. *Hydra*. Detroit, MI: KidHaven Press, 2009.
- McKerley, Jennifer Guess. *The Kraken*. Detroit, MI: KidHaven Press, 2008.
- Newquist, H. P. *Here There Be Monsters: The Legendary Kraken and the Giant Squid*. Boston, MA: Houghton Mifflin, 2010.

GLOSSARY

acid (AS-id) poison that burns through things
anchor (ANG-kur) to firmly stick to the ground
camouflage (KAM-uh-flahzh) to blend into one's surroundings
capsizes (KAP-size-iz) causes to turn over
fangs (FANGZ) sharp, pointy teeth
immortal (ih-MOR-tuhl) being able to live forever
maw (MAW) a beast's mouth and jaws
Norse (NORS) coming from the Norway area
opponents (uh-POH-nuhnts) fighters who compete against each other
regenerate (ree-JEN-uh-rate) to heal quickly and grow again
schools (SKOOLZ) groups of fish
serpent (SUR-puhnt) snake
tentacles (TEN-tuh-kuhlz) long, moving arms with suckers
underworld (UHN-dur-wurld) a place where dead souls live
venom (VEN-uhm) poison that needs to be injected
victor (VIK-tur) the winner
warriors (WOR-ee-urz) fighters
whirlpool (WURL-pool) a powerful pool of swirling waters

INDEX